Evidence-Based Practices for

Strategic and Tactical Firefighting

JONES & BARTLETT
LEARNING

World Headquarters
Jones & Bartlett Learning
5 Wall Street
Burlington, MA 01803
978-443-5000
info@jblearning.com
www.jblearning.com

Jones & Bartlett Learning books and products are available through most bookstores and online booksellers. To contact Jones & Bartlett Learning directly, call 800-832-0034, fax 978-443-8000, or visit our website, www.jblearning.com.

Substantial discounts on bulk quantities of Jones & Bartlett Learning publications are available to corporations, professional associations, and other qualified organizations. For details and specific discount information, contact the special sales department at Jones & Bartlett Learning via the above contact information or send an email to specialsales@jblearning.com.

Copyright © 2016 by Jones & Bartlett Learning, LLC, an Ascend Learning Company

All rights reserved. No part of the material protected by this copyright may be reproduced or utilized in any form, electronic or mechanical, including photocopying, recording, or by any information storage and retrieval system, without written permission from the copyright owner.

The content, statements, views, and opinions herein are the sole expression of the respective authors and not that of Jones & Bartlett Learning, LLC. Reference herein to any specific commercial product, process, or service by trade name, trademark, manufacturer, or otherwise does not constitute or imply its endorsement or recommendation by Jones & Bartlett Learning, LLC and such reference shall not be used for advertising or product endorsement purposes. All trademarks displayed are the trademarks of the parties noted herein. *Evidence-Based Practices for Strategic and Tactical Firefighting* is an independent publication and has not been authorized, sponsored, or otherwise approved by the owners of the trademarks or service marks referenced in this product.

There may be images in this book that feature models; these models do not necessarily endorse, represent, or participate in the activities represented in the images. Any screenshots in this product are for educational and instructive purposes only. Any individuals and scenarios featured in the case studies throughout this product may be real or fictitious, but are used for instructional purposes only.

The National Fire Protection Association, the International Association of Fire Chiefs, the National Institute of Standards and Technology, the Underwriters Laboratories, and the publisher have made every effort to ensure that contributors to *Evidence-Based Practices for Strategic and Tactical Firefighting* materials are knowledgeable authorities in their fields. Readers are nevertheless advised that the statements and opinions are provided as guidelines and should not be construed as official the National Fire Protection Association, the International Association of Fire Chiefs, the National Institute of Standards and Technology, and the Underwriters Laboratories policy. The recommendations in this publication or the accompanying resource manual do not indicate an exclusive course of treatment. Variations taking into account the individual circumstances, nature of medical oversight, and local protocols may be appropriate. The National Fire Protection Association, the International Association of Fire Chiefs, the National Institute of Standards and Technology, the Underwriters Laboratories, and the publisher disclaim any liability or responsibility for the consequences of any action taken in reliance on these statements or opinions.

Production Credits

Chief Executive Officer: Ty Field
President: James Homer
Chief Product Officer: Eduardo Moura
Vice President, Publisher: Kimberly Brophy
Director of Sales, Public Safety Group: Patricia Einstein
Executive Editor: Bill Larkin
Senior Development Editor: Alison Lozeau
Associate Director of Production: Jenny L. Corriveau
Associate Production Editor: Nora Menzi

Senior Marketing Manager: Brian Rooney
Art Development Editor: Joanna Lundeen
VP, Manufacturing and Inventory Control: Therese Connell
Composition: Cenveo Publisher Services
Cover Design: Kristin E. Parker
Rights and Photo Research Coordinator: Ashley Dos Santos
Cover Image: Courtesy of NIST.
Printing and Binding: Courier Companies
Cover Printing: Courier Companies

Author: David Schottke

Library of Congress Cataloging-in-Publication Data

Evidence-based practices for strategic and tactical firefighting / International Association of Fire Chiefs, National Fire Protection Association.
    pages cm
  ISBN 978-1-284-08410-8
1. Fire extinction.   2. Fire prevention.   I. International Association of Fire Chiefs.   II. National Fire Protection Association.
  TH9130.E95 2016
  628.9'25—dc23
                                    2014039073

6048
Printed in the United States of America
18 17 16 15 14   10 9 8 7 6 5 4 3 2 1

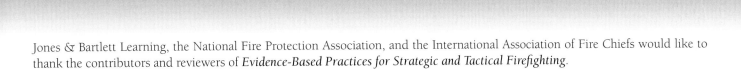

Acknowledgments

Jones & Bartlett Learning, the National Fire Protection Association, and the International Association of Fire Chiefs would like to thank the contributors and reviewers of *Evidence-Based Practices for Strategic and Tactical Firefighting*.

Author

David Schottke
Fairfax Volunteer Fire Department
Fairfax, Virginia

Contributors and Reviewers

Greg Barton
Deputy Fire Chief, Beverly Hills Fire Department
Beverly Hills, California

Jeff Elliott
Fire Service Program Director, Tennessee Fire Service & Codes Academy
Bell Buckle, Tennessee

William Guindon
Director, Maine Fire Service Institute
Brunswick, Maine

Roger C. Hawks
Executive Director, Tennessee Fire Service & Codes Academy
Bell Buckle, Tennessee

Shawn Kelley
Special Assistant to the Executive Director, International Association of Fire Chiefs (IAFC)
Fairfax, Virginia

Stephen Kerber
Director, Underwriters Laboratories (UL) Firefighter Safety Research Institute
Northbrook, Illinois

Daniel Madrzykowski
Fire Protection Engineer, National Institute of Standards and Technology (NIST), U.S. Department of Commerce
Gaithersburg, Maryland

Tony Mecham
Fire Chief, CAL FIRE and the San Diego County Fire Department
San Diego, California

Dan Munsey
Division Chief, San Bernardino County Fire Department
San Bernardino, California

Derek Parker
Captain, Sacramento Fire Department
Sacramento, California

Phillip Russell
Chief of Training and Administration, South Carolina Fire Academy
Columbia, South Carolina

John Shafer
Green Maltese LLC
Captain, Greencastle Fire Department
Greencastle, Indiana

Peter Silva, Jr.
Fire Service Education Director, Wisconsin Technical College System
Madison, Wisconsin

Jason Sparks
Regional Coordinator, Tennessee Fire Service & Codes Academy
Bell Buckle, Tennessee

Peter Van Dorpe
Assistant Chief, Algonquin-Lake in the Hills Fire Protection District
Lake in the Hills, Illinois

Michael Ward
Senior Associate, Fitch and Associates
Platte City, Missouri

Evidence-Based Practices for Strategic and Tactical Firefighting

Introduction

This is an exciting time to be a fire fighter. Today we are experiencing major changes in our understanding of basic fire behavior and the tactics used to attack fires. If we do not understand the impact of these changes, we may put ourselves in positions that are so unsafe that they prevent us from accomplishing our mission of preserving life and property.

Changes are occurring in building materials and construction techniques used to construct today's buildings. Household furnishings and other building components are increasingly constructed from petroleum-based materials. Today's fires release energy faster, reach flashover potential sooner, may reach higher temperatures, and are much more likely to become ventilation-limited than did building fires of even a few years ago. Because of these differences, it is important that our techniques for ventilating, applying water, and performing rescue reflect the changes in our modern building construction.

Historically, the principal means of developing firefighting strategy and tactics has been based on the observations and experiences of fire fighters. These observations and experiences provided us with valuable tools to more effectively fight fires. Yet, experience-based techniques did not provide us with the means to fully measure and understand the actual progression of fire and the impact of each action we take at a fire scene. Because of these limits, we have sometimes drawn inaccurate conclusions. These conclusions have sometimes resulted in ineffective and counterproductive courses of action.

One primary reason for the increase in our understanding of fire behavior is the result of live fire experiments. Over the last 15 years, the National Institute of Standards and Technology (NIST) and the Underwriters Laboratories (UL) have conducted controlled <u>fire scenarios</u> in specially constructed laboratories and in actual houses **FIGURE 1**. The researchers used a variety of instruments to measure the temperatures throughout the structure at various heights within each room, the heat release rates and heat generated from the room contents, the airspeed and direction in and out of the fire compartment, and—from ventilation points—the visibility and the chemical makeup of the fire gases within and outside the structure. All these observations and data are carefully recorded so they can be completely analyzed. Each change that is made to the fire scene, such as ventilating or applying water, can be measured. Because the researchers have been able to conduct these experiments multiple times, it is possible to determine how each action taken by fire fighters impacts the growth and extinguishment of the fire.

Because these changes have occurred so quickly, it is hard for instructional materials to keep current. This resource supplement to *Fundamentals of Fire Fighter Skills, Third Edition*, is created to introduce you to some of the findings and conclusions gained from these experiments. It is not designed to supply all the information you need to implement changes in suppression techniques, but rather to raise your awareness of these changes and to guide you to current resources and future changes in instructional materials. This additional resource supplement describes some of the advances in our knowledge of the principles of fire behavior and compares building techniques used in older or legacy construction with modern construction techniques used in residential construction today. Because current construction techniques, building materials, and furnishings differ from those used in older buildings, the progression of today's fires is different. These changes require modification of our approach to fighting fires today.

FIRE FIGHTER Tips

Because of increased fuel loads and modern lightweight construction techniques, today's fires grow faster and release more heat. This has led to greater adoption of residential sprinkler systems. A house with working smoke detectors and a residential sprinkler system increases the occupants' chance of surviving a fire by about 80%.

FIGURE 1 A live fire experiment conducted by NIST and UL with the Fire Department of New York (FDNY) on Governer's Island, NY.
Courtesy of NIST.

Fire Behavior

Firefighting can be broken into many steps. Here we consider changes in modern fire behavior and the importance of controlling ventilation. Offensive fire attack and changes in water application are also described, and we will explore improved tactics when searching for fire victims. By exploring these topics, we can improve fire fighter safety and survivability of building occupants and improve the effectiveness of fire suppression.

For as long as there has been fire, it has followed the same rules of physics and chemistry. A fire is produced when fuel and oxygen are combined with a source of ignition, and today's fires follow these rules as they did in the earliest of times. A simple example would be a campfire, which combines wood and oxygen with a source of ignition to produce a fire. Left unattended, this fire will start with an incipient stage, grow to a fully developed fire, and then proceed to a decay stage when the available fuel is exhausted **FIGURE 2**. We have all learned these classic stages of fire growth. While this sequence of fire growth holds true for campfires, fires in recently constructed buildings do not necessarily follow these stages. In recent years, fire fighters have encountered a surprising increase in the number of violent and rapid flashovers at residential fires. Fire fighters have not completely understood the conditions that cause these flashovers. Over the last several years, with support from the Assistance to Firefighters Grants and the Fire Prevention and Safety Grants programs, NIST and UL have conducted an extensive series of experiments specifically designed to provide the fire service with the knowledge and tools it needs to better understand this phenomenon **FIGURE 3**. These experiments included component testing of modern furnishings, furnished room experiments, full-scale house burns in the laboratory, and full-scale burns in acquired structures. This evidence-based research points to topics that the fire service needs a better understanding of in order to operate safely in today's fire environment, including:

- Fire dynamics
- Ventilation and fire flow paths: how fires spread
- Applying water: coordinating fire attack
- Rescue and safety considerations
- Basement fires

■ Fire Dynamics

The interior structures of older houses were built primarily of wood products and plaster. The exteriors were wood, brick, and, in some cases, asbestos. They were finished in wood trim and furnished with contents that were made primarily of natural fibers such as wool and cotton. These houses were built using solid wood beams with wooden siding or solid masonry walls. The homes had limited insulation in the walls

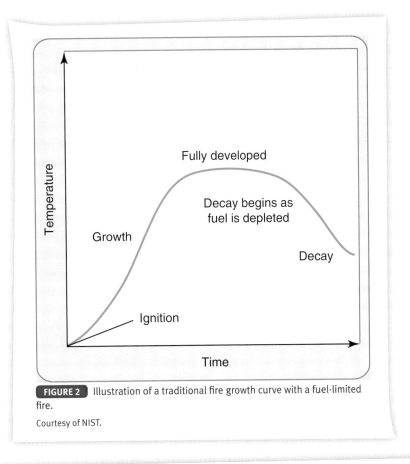

FIGURE 2 Illustration of a traditional fire growth curve with a fuel-limited fire.
Courtesy of NIST.

FIGURE 3 The legacy room with older furnishings has little flame showing 3 minutes and 25 seconds after ignition, while the modern room furnished with petroleum-based products has flashed over in the same time period.
Courtesy of UL.

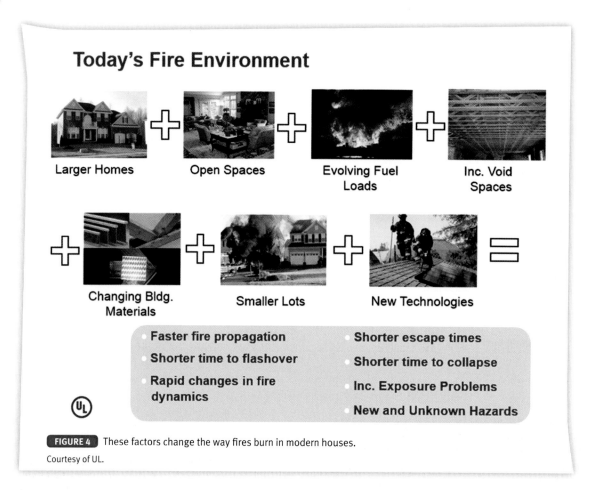

FIGURE 4 These factors change the way fires burn in modern houses.
Courtesy of UL.

and ceilings, and windows had only a single pane of glass. This type of older construction is called legacy construction. For years, fire fighters have fought fires in these types of buildings. Fires in this type of structure were often fought using an aggressive interior attack.

Houses built since the mid-1980s are constructed using different construction techniques and materials. Solid dimensional lumber has been replaced with manufactured wood beams or trusses made from as little wood as possible or from light-gauge steel. Houses are finished with a wide variety of plastic-based materials derived from petroleum products, such as polyurethanes and polystyrenes, and filled with petroleum-based foam-filled furniture. Furniture constructed from plastic-based materials is found in buildings of all types of construction and contributes to a greater fuel load and a more rapidly developing fire. Modern houses are larger, have more open spaces, and are constructed to be energy-efficient. The result is a house that is more tightly sealed. Windows are often constructed with double panes, which means that they often require more energy from a fire before they will fail—modern windows do not break as quickly in a fire. These changes in modern home construction and furnishings greatly impact the growth, progression, ventilation, and suppression of fires in modern construction **FIGURE 4**.

NIST and UL have conducted fire experiments comparing the growth of a fire in a room with synthetic furnishings versus one with older contents. In these experiments, a room with legacy furnishings took a longer time to flashover or did not reach flashover. A room with synthetic furnishings, however, flashed over in a very short period of time. In other words, a room with furnishings produced from natural materials burns much more slowly than a room furnished with synthetic or plastic-based materials. Perhaps the greatest value of these experiments is their ability to show that fires fueled by synthetic furnishings contain a much greater fuel load, progress faster, and flashover more quickly. This creates a much more dangerous condition for both building occupants and fire fighters.

Experiments have also documented that a room-and-contents fire in a modern house will often flashover and then enter a dormant phase due to a decreased concentration of oxygen in the house. Research has also documented that a typical room-and-contents fire in a standard residential occupancy may use up the available oxygen before it reaches flashover. This produces a ventilation-limited fire where the fire goes into a dormant state or decay stage. Often no flame is visible, and the fire appears to be no longer actively burning. In this decay stage, the fire growth and development is limited only by a lack of available oxygen. However, there is still a substantial amount of heat in the space and the heated contents continue to pyrolize, sending additional fuel into the space in the form of smoke and fire gases. There is a huge amount of energy in the form of smoke and superheated gases, and all that is lacking is oxygen. To fire fighters, a fire in the decay stage will look like it has gone out or that it is very small. Experiments have repeatedly shown, however, that when oxygen, in the form of outside air, is introduced into a <u>ventilation-limited</u> fire compartment

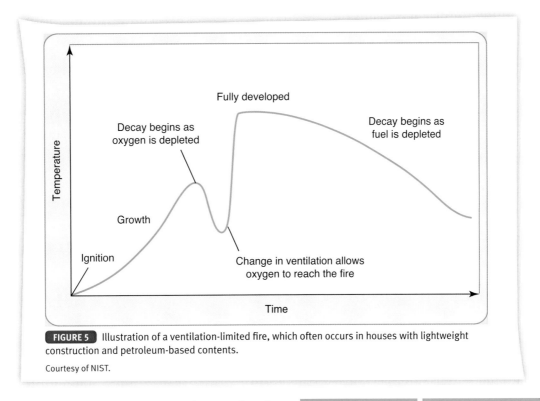

FIGURE 5 Illustration of a ventilation-limited fire, which often occurs in houses with lightweight construction and petroleum-based contents.
Courtesy of NIST.

in the decay stage, it can rapidly generate a ventilation-induced flashover in a short period of time **FIGURE 5**.

Research indicates that fires in modern residential occupancies are likely to become ventilation-limited prior to the arrival of the first due fire company. This means that introduction of air will result in rapid fire growth. The research has also demonstrated that fire fighters making entry through a front door can introduce enough air into the fire area to produce rapid fire growth and flashover. Fire fighters need to carefully consider what constitutes ventilation. Opening any door, window, skylight, or roof introduces oxygen into a burning building. The fire service has not traditionally taught that making entry into a building is part of ventilation, yet studies have shown that opening the front door has a profound impact on the growth of the fire. Repeated experiments produced a violent flashover shortly after the front door is opened. This action must be considered to be a part of ventilation.

This research has also demonstrated that our traditional assumptions about the effects of ventilation need to be modified. Most fire service texts describe ventilation as the systematic removal of heat, smoke, and fire gases from the building, and replacing them with cool fresh air. While this definition is and always has been technically true, it can lead to a misapplication of the principle. What the research has repeatedly demonstrated is that when dealing with ventilation-limited fires, any cooling effect produced by ventilation alone is minimal and very short-lived. Because of the extremely fuel-rich environment found on today's fireground, ventilation that is not preceded by, concurrent with, or immediately followed by effective suppression will introduce enough oxygen to rapidly bring the fire area to flashover. **TABLE 1** shows the conclusions for residential fire behavior that were drawn from the fire experiments conducted by NIST, UL, and the Fire Department of New York (FDNY).

TABLE 1	Residential Fire Behavior

- Increasing the air flow to a ventilation-limited structure fire by opening doors, windows, or roof openings will increase the hazard from the fire for both building occupants and fire fighters.
- Increasing the air flow to a ventilation-limited structure may lead to a rapid transition to flashover.
- There is a need for a coordinated fire attack to coordinate the activities of size-up, entry, ventilation, search and rescue, and application of water.

Courtesy of NIST.

Fire Fighter Key Points

Some important points to remember regarding fire dynamics are:

- The stages of fire development change when a fire becomes ventilation-limited; it is common with today's fire environment to have a decay period prior to flashover, which emphasizes the importance of ventilation and its timing.
- The absence of visible smoke means nothing! A common event noted during the experiments was that once the fire became ventilation-limited, the smoke being forced out of the gaps of the houses greatly diminished or stopped altogether. No smoke showing during the size-up should increase awareness of the potential conditions inside.
- Structural collapse should always be considered in your size-up. All residential floors can collapse in your operational time frame, especially with an unprotected engineered floor system.

© Photos.com

■ Ventilation and Fire Flow Paths: How Fires Spread

In order to most effectively extinguish fires, we need to understand how they spread. All fire fighters have learned about fire spread based on the principles of conduction, convection, and radiation, and the classic fire triangle is still helpful to understand the growth and suppression of fires. Content fires generate large volumes of smoke, particulates, and gases, all of which are considered fuel. If this fuel is hot enough, it will ignite with the introduction of adequate oxygen.

In addition to these principles of fire spread, we need to add another factor that has not gotten the attention it deserves. Fire spread is largely a pressure-driven phenomenon. Fires spread along a flow path. A flow path is the lower pressure space between an inlet for fresh air (such as an open door or window), the fire, and the higher pressure space between the fire and the outlet for hot gases and smoke, such as an open window or roof opening. Fires in imperfectly sealed buildings produce pressures that, while very small, still result in pressure differentials that cause significant movement of fire gases throughout the building and through external openings **FIGURE 6** . Understanding how pressure differentials and fire flow paths influence the growth of fires leads us to create more effective ventilation and fire control techniques. Put simply, while we cannot always say that heat always rises or that fire gases will always move upward, we can say

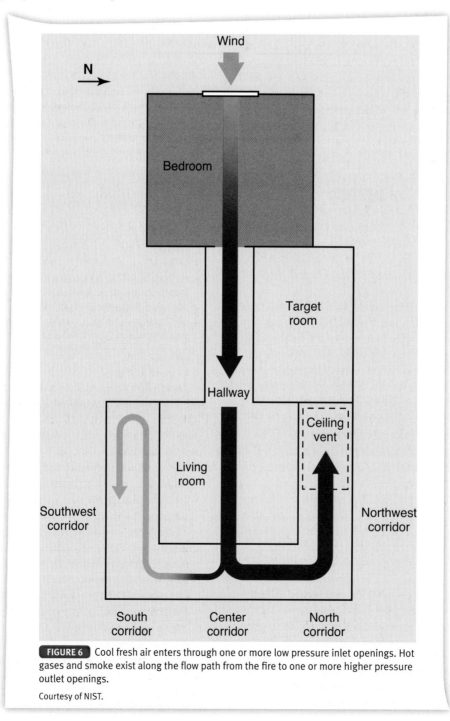

FIGURE 6 Cool fresh air enters through one or more low pressure inlet openings. Hot gases and smoke exist along the flow path from the fire to one or more higher pressure outlet openings.

Courtesy of NIST.

that high pressure will always move toward low pressure. Furthermore, hot gases moving away from the seat of the fire create a partial vacuum behind them that draws any available cooler fresh air toward the seat of the fire, accelerating its growth. In a fire where only the front door is open, the fire will create a flow path of hot gases exiting through the top part of the door while the bottom half of the door will allow the introduction of cool oxygen-rich air into the fire. This type of opening produces a two-way or bidirectional flow path through the front door. Some fire flow paths move in a single direction. Single-direction flow paths either supply oxygen by creating an inlet path or release hot gases by creating an exhaust flow path. Single direction flow paths can occur when there are multiple openings, such as an open front door and an open window, to carry gases in a single direction. Fresh air enters through the front door, and hot gases and smoke exit through the open window. The hot gases are pushed along the flow path. If the flow path moves through the building before exiting to the outside and adequate oxygen mixes with the hot gases, the entire flow path can become a rapidly moving wall of hot gases or flame. Even a fully encapsulated fire fighter cannot survive these conditions for more than a few seconds.

Fire growth and development are impacted by both the amount of air entering the fire area and by the amount of fire gases exiting the fire area. Opening a door or window provides oxygen to a ventilation-limited fire. Providing an opening in the roof creates an exit flow path for hot fuel and will accelerate air flow into the fire as well as spread the fire along the flow path. This significant increase in oxygen combining with the hot, fuel-rich environment of a ventilation-limited fire can result in explosive fire growth FIGURE 7.

Wind is a powerful force in influencing the direction of a flow path. Any time a window or door is opened on the side of a building that faces the wind—the windward side—a huge amount of oxygen will be introduced into the fire. This action may be accidental, such as a glass window or door breaking because of heat from the fire, or it can be intentional, such as breaking a window or door for ventilation. This sudden increase in oxygen combined with hot flammable gas from the fire can result in rapid fire growth but can also produce a sudden change in the direction of the flow path due to the pressure imposed by the wind. Think of the wind as a giant positive pressure fan forcing huge quantities of oxygen into a ventilation-limited fire. Remember to keep the wind at your back during a fire attack. The National Fire Protection Association (NFPA) Fire Protection Research Foundation has conducted experiments on wind-driven fires. More information about these studies is available at the NFPA website, Fire Fighting Tactics Under Wind Driven Conditions.

One means of controlling a fire is to control the amount of oxygen that is available, and understanding what a flow path is will aid in that control. Limiting the oxygen available to the fire until other suppression activities can be implemented helps keep a fire smaller. Because opening the front door of a house supplies the fire with oxygen and may quickly produce a violent flashover, ventilation needs to be considered as a critical part of our fire suppression activities, and it must always be conducted in coordination with, and only when, effective suppression efforts can be established and sustained FIGURE 8. Leaving the front door closed as long as possible aids significantly in controlling the fire by limiting the amount of oxygen supplied to the fire, as does stationing a fire fighter at the front door to keep the door partly closed as the hoseline is advanced into the fire. The conclusions that were drawn regarding the fire flow path from the fire experiments conducted by NIST, UL, and the FDNY are shown in TABLE 2.

FIRE FIGHTER Tips

Think of the fire flow path as being a river bed through which water flows. The location of the river bed flow path is dependent on the location of air entering the fire compartment and the location of air exiting the fire compartment. As you create openings into the fire compartment, you influence the location and direction of the flow path by opening doors, breaking windows, or creating ventilation openings in the roof. The flow of air will be cooler on the inlet side of the riverbed flow path and hotter on the outlet side of the riverbed flow path.

Fire Fighter Key Points

Some important points to remember regarding ventilation and fire flow paths are:

- Forcing the front door is ventilation and must be thought of as ventilation. While forcing entry is necessary to fight the fire, it must also trigger the thought that air is being fed to the fire and the clock is ticking before either the fire gets extinguished or it grows until an untenable condition exists, jeopardizing the safety of everyone in the structure.
- Once the front door is opened, attention should be given to the flow through the front door. A rapid rush of air or a tunneling effect could indicate a ventilation-limited fire.
- Every new ventilation opening provides a new flow path to the fire and vice versa. This could create very dangerous conditions when there is a ventilation-limited fire. You never want to be between where the fire is and where it wants to go without water or a door to close.
- Fire showing does not mean the fire is vented—it means it is venting and additional ventilation points will grow the fire if water is not applied.
- "Taking the lid off" does not guarantee positive results. Vertical ventilation is the most efficient type of natural ventilation. It allows for the most hot gases to exit the structure; however, it also allows the most air to be entrained into the structure. Coordination of vertical ventilation must occur with fire attack—just like with horizontal ventilation.
- Interior fire attack is still the most common and important fire attack method, but emphasis needs to be put on controlling ventilation and cooling fire gases. Apply water to the smoke to cool it as you move through it.

© Photos.com

 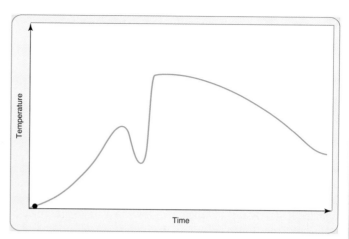

A

 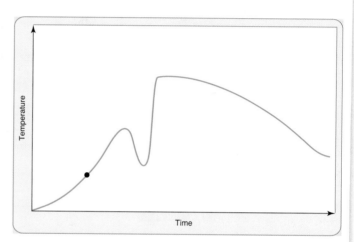

B

 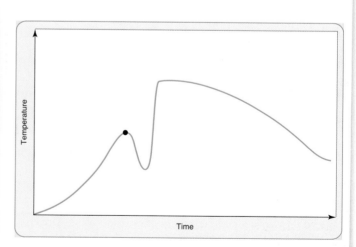

C

FIGURE 7 A ventilation-limited fire sequence, in which the main picture in each panel shows the room where the fire was started with a small picture on the lower right showing the front of the fire building at that time. The graph to the right shows the position of the fire on the fire growth curve. **A.** The initial flaming and growth of the fire. **B.** Halfway through the initial growth curve. **C.** Fire at its initial peak. (*continues*)

Courtesy of NIST.

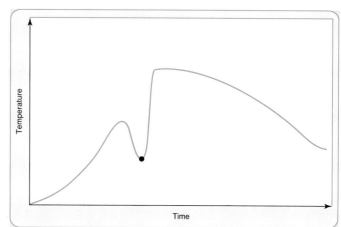

D

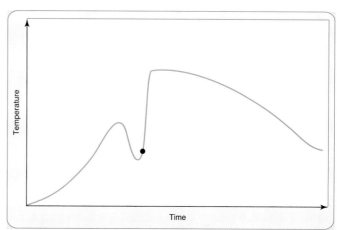

E

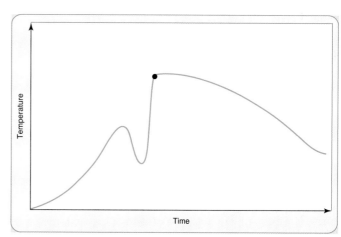

F

FIGURE 7 A ventilation-limited fire sequence, in which the main picture in each panel shows the room where the fire was started with a small picture on the lower right showing the front of the fire building at that time. The graph to the right shows the position of the fire on the fire growth curve. **D.** Fire in the decay stage just before the fire becomes ventilation-limited. **E.** The fire re-ignites 5 seconds after the front door is opened, increasing the oxygen supply to the fire. **F.** One minute after the front door is opened, the room is fully involved with a flashover.

Courtesy of NIST.

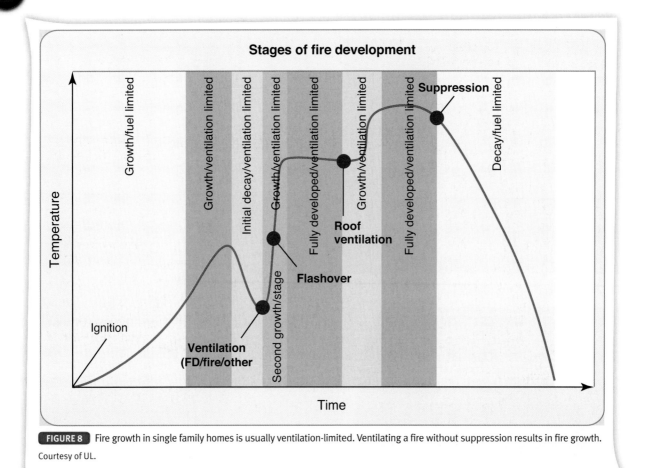

FIGURE 8 Fire growth in single family homes is usually ventilation-limited. Ventilating a fire without suppression results in fire growth.
Courtesy of UL.

TABLE 2	Fire Flow Path

- Improving the inlet and/or exhaust paths from the seat of the fire will result in fire growth and spread.
- Interrupting the fire flow path by limiting or controlling the inlet or controlling the outlet can limit fire growth.
- Controlling the door—keeping doors closed allows less oxygen into the fire and equals lower temperatures.
- Anyone in the exhaust portion of the flow path—between the fire and the direction of its travel—is in a high hazard location.
- Controlling the flow path improves victim survivability.

Courtesy of NIST.

■ Applying Water: Coordinating Fire Attack

Our primary means of suppressing fires is to apply water. The research conducted by NIST and UL provide valuable knowledge about more effective means of applying water to a fire. It has been widely accepted that attacking a fire from the outside may push the fire through other parts of the building. Traditionally, it has been taught that a residential fire should be attacked through the front door or from the unburned side. This has often led to the assumption that offensive fire attacks must be made exclusively from the inside of the structure and that the fire must always be approached from the unburned side, effectively "pushing the fire" away from the uninvolved portion of the building and from any occupants that might be behind the advancing hose team. This technique requires fire fighters to enter a burning building, often with low visibility and little idea of where the fire is located. NIST and UL research demonstrates that these assumptions are not true.

As NIST and UL measured the movement of the fire, they determined that in no case was it possible to push the fire with a stream of water. Hose streams do not push fire or fire gases. They determined that the movement of the fire is dependent on the flow path of heated smoke and gases. The fire flow path carries the fire from an area of high pressure to areas of lower pressure. After verifying these results with multiple experiments, they determined that if water was introduced into the fire from the outside, it produced amazing cooling, not just in the fire compartment, but also in other areas of the fire building. Significant cooling was achieved even when the water entered the flow path at a distance from the seat of the fire. In these experiments, a straight stream of water flowing at 180 gallons of water per minute (gpm) (11 liters per second) was applied for 28 seconds and flowed 84 gallons (318 liters) of water into the house at some distance from the fire. This water reduced the temperature in the front of the living room from 1200°F (649°C) to 300°F (149°C) (**Note**: These numbers apply only to the Modern Fire Behavior experiments conducted at Governor's Island by FDNY, NIST, and UL.) **FIGURE 9**. These experiments showed that this offensive exterior attack—introducing water from the outside—reduced temperatures in other parts of the house at some distance

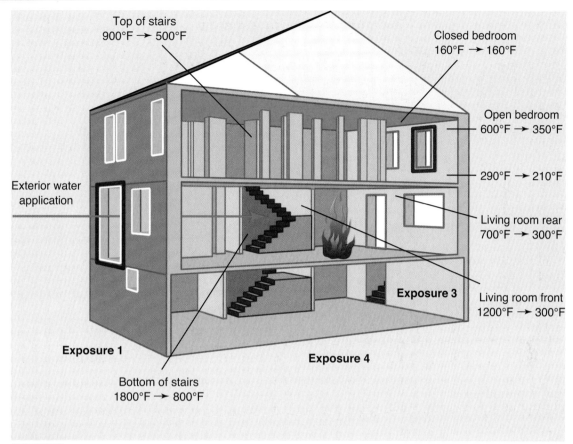

FIGURE 9 An offensive exterior attack some distance from the fire reduces temperatures throughout the house and partially suppresses the fire. The temperatures listed are from before and after the exterior application of water.
Courtesy of NIST.

from the fire, but did not completely extinguish the fire. It slowed the growth of the fire by cooling huge quantities of very hot gaseous fuel and solid fuel below its <u>ignition temperature</u>. This <u>offensive exterior attack</u> is sometimes referred to as a <u>blitz attack</u>, a <u>transitional attack</u>, or, by the using a military metaphor, <u>softening the target</u>.

The effect of this offensive exterior attack is significant but short-lived. In order to prevent the fire from regaining its former state, to complete extinguishment, and to search for and remove any occupants, it is necessary to enter the building and to aggressively move onto the remaining fire immediately following the knock down from the exterior. This combined attack is safer for fire fighters because when entering the building, the internal temperature is cooler, visibility is increased, and the threat of a flashover is greatly reduced. The transitional attack can often put effective water on the seat of the fire sooner, thus making conditions better throughout the building for any trapped occupants. With the fire knocked down, more aggressive ventilation can follow without risk of inducing flashover. Visibility improves and fire fighters can move through the building more effectively, efficiently, and safely, while conducting rescue operations and other tasks. The results of these experiments require us to reconsider our long-standing approaches to ventilation and the application of water into the fire buildings **FIGURE 10**. **TABLE 3** shows the conclusions on exterior fire attacks that were drawn from the fire experiments conducted by NIST, UL, and the FDNY.

Fire Fighter Key Points

Some important points to remember regarding applying water/coordinating fire attack are:
- You cannot push fire with water. Air entrained by a stream will cause heat to flow through a flow path. You need to reduce the amount of air introduced with the stream of water and let the hot gases flow out as you are applying the water through an opening. Just like any other tactic, there is a correct way to flow water into an opening, such as a window or door, if you do not want heat to follow the flow path.
- It is not possible to make statements about the effectiveness of ventilation unless one includes timing. Venting does not always lead to cooling; well-timed and coordinated ventilation leads to improved conditions.
- Effective firefighting depends on a coordinated fire attack. Coordinating entry with ventilation, isolation, suppression, and rescue is of paramount importance.

© Photos.com

FIGURE 10 Fire fighters using a transitional attack to reduce the burning and cool the interior before entering the burning building. **A.** The second story window. **B.** Exterior attack.

Courtesy of UL.

TABLE 3	Exterior Fire Attack

- An offensive exterior fire attack through a window or door, even when it is the only exterior vent, will not push fire.
- Water application is most effective if a straight stream is aimed through the smoke into the ceiling of the fire compartment. Water should be flowed for about 10–20 seconds. This technique allows heated gases to continue to vent from the fire compartment while cooling the hot fuel inside. Fog patterns should not be used in this application. The fog pattern entrains large volumes of air and pushes air into the building. A fog stream can also block a ventilation opening, effectively changing the fire flow path.
- Applying a hose stream through a window or door into a room involved in a fire resulted in improved conditions throughout the structure.
- Even in cases where the front and rear doors were open and windows had been vented, application of water through one of the vents improved conditions throughout the structure.
- Applying water directly into the compartment as soon as possible resulted in the most effective means of suppressing the fire.
- Transitional attack is an offensive exterior fire attack that occurs just prior to entry, search, and tactical ventilation. This technique is also known as a blitz attack, a transitional attack, or softening the target.
- The transitional attack should begin from the outside, but it is necessary to finish it from the inside.
- Coordinate the fire attack with vertical ventilation—do not ventilate before an attack stream is ready.

Courtesy of NIST.

■ Rescue and Safety Considerations

The same research that increased our understanding of fire flow paths, the effects of ventilation in ventilation-limited fires, and the use of a transitional fire attack also increases our knowledge of rescue techniques. Anytime you can close a door between the fire and other parts of the building, you increase the chance for survival of any trapped fire fighters or building occupants. Isolating a room from the fire can provide a safe environment for a longer period of time. The rescue technique of vent–enter–search (VES), requiring a fire fighter to ventilate a bedroom window from the outside, enter a bedroom through the open widow, and search for victims, is a technique that creates a dangerous situation for the fire fighter and for any victims in that room. By opening the bedroom window, you can provide a low pressure vent for the fire to move toward. Recent experiments have demonstrated that this rescue technique is much safer if the interior door is closed prior to conducting the search. This isolates this room from the fire flow path. The modified VES process becomes vent–enter–isolate–search (VEIS) **FIGURE 11**.

VEIS consists of four steps. First, the fire fighter identifies the location of possible fire victims in a burning building. For example, a bedroom in the middle of the night is likely to contain a fire victim. This room is quickly ventilated by breaking out a window or by other means. Then the fire fighter quickly enters the room and locates the door into that room and immediately closes it. This isolates this room from the main body of the fire and decreases the chance that it becomes part of the fire flow path. The fire fighter then systematically

FIGURE 11 A VEIS rescue may be lifesaving at this fire.
Courtesy of UL.

searches the room for victims. This change in the search technique increases the safety of both fire fighters and fire victims. Conclusions from the fire experiments conducted by NIST, UL, and the FDNY regarding the impact on building occupants are shown in **TABLE 4**.

TABLE 4	Impact on Building Occupants

- Suppressing the fire from the exterior as soon as possible improves potential survival time.
- Additional ventilation that is not immediately followed by effective fire suppression reduced potential survival time.
- Being in the exhaust flow path of the fire resulted in reduced potential survival time.
- Controlling the flow path improves victim survivability.
- Controlling the door to a room when performing VEIS improves the safety of the fire fighter and the building occupant.
- Compartmentation (being behind a closed interior door) prior to fire department arrival provided increased protection compared to being in a room or area connected to the fire.
- Greater distance from the fire improved chances of survival.

Courtesy of NIST.

Fire Fighter Key Points

Some important points to remember regarding rescue and safety considerations are:
- If you add air to the fire and do not apply water in the appropriate time frame, the fire gets larger and safety decreases. Coordination of the fire attack crew is essential for a positive outcome in today's fire environment.
- The greatest probability of finding a victim is often behind a closed door.
- During a VEIS operation, primary importance should be given to isolating the room being searched by closing the door to the room. This eliminates the impact of the open vent and increases tenability for potential occupants and fire fighters while the smoke ventilates from the now isolated room.
- If you get into trouble and need to escape, closing a door between you and the fire will buy you valuable time.
- Floor sag is a poor indicator of floor collapse. It may be very difficult to determine the amount of deflection while moving through a structure.
- Sounding the floor for stability is not reliable and therefore should be combined with other tactics to increase safety.

© Photos.com

Basement Fires

Research involving basement fires has also provided valuable information. Many vertical voids within a building originate or terminate in the basement, providing ample opportunities for fire gases to spread throughout the building. In many parts of the country, especially in newer subdivisions with lightweight construction, basement ceilings are left unfinished when the home is sold. Under these circumstances, fires originating in the basement will quickly involve the floor system, resulting in a failure of the floor over the fire and early collapse, with the potential for causing injury or death to fire fighters entering the building with conditions of limited visibility. The traditional tactic of pushing down the interior basement stairs to the seat of the fire should no longer be considered a safe option.

One of the findings of recent research on basement fires was that water applied via the interior basement stairs had a limited effect on cooling the basement or extinguishing the fire. However, water applied through an external window or door quickly darkened down the fire and reduced temperatures throughout the building, and no fire or hot gases were "pushed" up the interior stairs **FIGURE 12**. This effect lasted for several minutes before the fire grew back to the size it was before the application of water. **TABLE 5** shows the conclusions drawn from the fire experiments conducted by NIST, UL, and the FDNY regarding basement fires.

TABLE 5	Basement Fires

- Fire flows from basement fires developed in locations other than the stairs, as the floor assembly often failed close to the location where the fire started.
- Flowing water at the top of the interior stairs had limited impact on basement fires.
- Offensive exterior attack through a basement window was effective in cooling the fire compartment.
- Offensive fire attack through an exterior door was effective in cooling the fire compartment.

Courtesy of NIST.

Fire Fighter Key Points

Some important points to remember regarding basement fires are:

- Thermal imagers may help indicate there is a basement fire but cannot be used to assess structural integrity from above.
- Attacking a basement fire from a stairway places fire fighters in a high-risk location due to being in the flow path of hot gases flowing up the stairs and working over the fire on a flooring system that has the potential to collapse due to fire exposure.
- Coordinating ventilation is extremely important. Ventilating the basement creates a flow path up the stairs and out through the front door of the structure—almost doubling the speed of the hot gases and increasing temperatures of the gases to levels that could cause injury or death to a fully protected fire fighter.

© Photos.com

FIGURE 12 Fire fighters using an offensive exterior attack. **A.** Basement fire with flames showing at the basement window. **B.** Water applied through an external basement window darkens the fire and reduces temperatures throughout the building. Application of water will not push the fire into other parts of the building.

Courtesy of NIST.

Continued Research

Among the current research that is being conducted by UL and NIST in cooperation with other organizations are experiments studying the most effective means of applying water to a fire and studies examining the dynamics of attic fires. These research projects will provide valuable information about the growth of fires, fire flow paths, effective ventilation techniques, and optimal suppression techniques. The results of this research will be included in educational materials as soon as they are available.

Summary

With this resource supplement to *Fundamentals of Fire Fighter Skills, Third Edition*, we have attempted to illustrate some of the findings discovered as the result of an ongoing series of experiments conducted by NIST, UL, and FDNY. Additional experiments are being conducted this year and are planned for the future. This additional resource supplement is designed to provide up-to-date information that continues to change, and because of these rapid changes, it is not possible for fire textbooks to always contain the most current findings. In order to keep up to date, it is important to study the information on the NIST and UL websites. Additional information about these fire experiments can be obtained online at the NIST website by searching under Firefighting Technology and at the UL website by searching under Firefighter Safety Research Institute.

Wrap-Up

Hot Terms

<u>blitz attack</u> An aggressive fire attack that occurs just prior to entry, search, and tactical ventilation; also referred to as an offensive exterior attack, softening the target, or a transitional attack.

<u>conduction</u> Heat transfer to another body or within a body by direct contact.

<u>convection</u> Heat transfer by circulation within a medium such as a gas or a liquid.

<u>fire scenarios</u> A series of experiments designed to measure the characteristics of fire growth, progression, and extinguishment. These are conducted under carefully controlled and monitored conditions.

<u>fire triangle</u> A geometric shape used to depict the three components of which a fire is composed: fuel, oxygen, and heat.

<u>flashovers</u> A transition phase in the development of a compartment fire in which surfaces exposed to thermal radiation reach ignition temperature more or less simultaneously and fire spreads rapidly throughout the space, resulting in full room involvement or total involvement of the compartment or enclosed space.

<u>flow path</u> The volume of air or hot gases, smoke, and small particles between an inlet and outlet that allows the movement of heat and smoke from the higher pressure within the fire area toward the lower pressure areas accessible through doorways, window openings, and roof openings.

<u>ignition temperature</u> Minimum temperature a substance should attain to ignite under specific test conditions.

<u>offensive exterior attack</u> An aggressive attack that occurs just prior to entry, search, and tactical ventilation; also referred to as softening the target, a blitz attack, or a transitional attack.

<u>radiation</u> The emission and propagation of energy through matter or space by means of electromagnetic disturbances that display both wave-like and particle-like behavior.

<u>softening the target</u> An aggressive offensive exterior fire attack that occurs just prior to entry, search, and tactical ventilation; also referred to as a blitz attack, an offensive exterior attack, or a transitional attack.

<u>transitional attack</u> An aggressive offensive exterior fire attack that occurs just prior to entry, search, and tactical ventilation; also referred to as an offensive exterior attack, softening the target, or a blitz attack.

<u>vent–enter–isolate–search (VEIS)</u> A method of searching for fire victims that consists of ventilating an enclosed space, such as a bedroom, entering the room, closing the door to isolate the room from the fire, and then quickly searching for any possible victims.

<u>ventilation-limited</u> A fire in an enclosed building that is restricted because there is insufficient oxygen available for the fire to burn as rapidly as it would with an unlimited supply of oxygen. Increasing the supply of oxygen to a ventilation-limited fire may result in a rapid flashover.

References

National Fire Protection Association (NFPA). Fire Fighting Tactics Under Wind Driven Conditions. http://www.nfpa.org/research/fire-protection-research-foundation/reports-and-proceedings/for-emergency-responders/fireground-operations/fire-fighting-tactics-under-wind-driven-conditions. Accessed October 17, 2014.

National Institute of Standards and Technology (NIST) Engineering Laboratory. Fire on the Web. http://www.nist.gov/fire/. Accessed September 29, 2014.

Underwriters Laboratories (UL) Firefighter Safety Research Institute. http://ulfirefightersafety.com. Accessed June 20, 2014.